WELDING ONE – TEST BOOKLET

Contains Module Examinations / Answer Keys, Performance Profile Sheets, and Performance Accreditation Tasks

NATIONAL
CENTER FOR
CONSTRUCTION
EDUCATION AND
RESEARCH

STORE THIS TEST BOOKLET IN A SECURE AREA!

To enhance the security and flexibility of standardized craft training testing, the National Center for Construction Education and Research offers two testing formats for the written examination. Instructors may provide testing for their trainees in one or both of the following ways:

- *By using the reproducible masters included in this booklet;*
- *By using the Windows-based TestGen Software to reformat existing exams and/or create final exams.*

Using the Reproducible Masters

1. Masters for the module examinations are found on the following pages. The tests can be photocopied for your local use. ***The answer keys are included in this test booklet. Be sure to keep answer keys and tests in a secure area!***

2. You may also choose the option of administering tests on overhead transparencies and having trainees fill in an answer sheet. This method improves test security since no copies of the tests are distributed.

Using the TestGen Software

The TestGen Software for Welding Level One/AWS Entry Level Welder – Phase One is available from Prentice Hall. Detailed instructions for installing and using the software are included with the software program. The software includes all of the module examinations questions found in this test booklet and also allows instructors to:

- Scramble test questions.
- Add test questions.*
- Merge module examinations to create a final exam.

To order the Welding Level One/AWS Entry Level Welder – Phase One TestGen software for only $30.00, call Prentice Hall at (800) 922-0579 and ask for ISBN# 0-13-103315-8.

** To remain in accordance with the NCCER's accreditation policies, instructors must test trainees with NCCER's module examination. Instructors may add questions to or supplement the module exam, but **they may not delete existing test questions**. Doing so represents a departure from NCCER Accreditation Policies and Guidelines and may seriously jeopardize either the accreditation status of the training program sponsor and/or any recognition of trainees, instructors, and trainers through the NCCER National Registry.*

NOTE ON PERFORMANCE TESTING

Performance Profile Sheet(s) are included in this test booklet in a format that can be easily photo-copied for each trainee. This examination is designed to measure competency in the tasks taught in each module.

Please note the number of tasks to be tested while teaching each module. Each trainee should be tested on all the tasks listed on the Performance Profile Sheet(s). Before the performance testing, the instructor should brief the trainees on:

- Test objectives and criteria.

- Safety precautions.

- Procedures for each task to be tested.

The instructor administering the performance testing should also do the following:

- Ensure that all of the needed equipment is available and operating properly.

- Set up the testing stations.

- Organize and administer the test in a way that allows for optimal performance.

- Complete the Performance Profile Sheet(s) for each trainee by assigning a pass/fail score for each listed task.

- Monitor adherence to all safety regulations and precautions.

- Provide adequate supervision to prevent injuries.

- Take immediate and effective action to remedy any emergency.

Accreditation Testing

If Performance Testing is done as part of the National Center for Construction Education and Research Standardized Craft Training Program, the following conditions must be met:

1. The Craft Instructor must hold valid NCCER instructor certification.

2. The training must be delivered through a Training Program Sponsor recognized by the NCCER.

3. For every module, the specific performance testing must be completed to the satisfaction of the instructor.

4. The results of the testing must be recorded on the Craft Training Report Form 200. This form must be provided to the local Training Program Sponsor to be forwarded to the NCCER National Craft Training Registry.

Welding One Test Booklet

Contains Module Examinations / Answer Keys and Performance Profile Sheets for the following Welding Level One Modules:

Module 29101-03   Welding Safety

Module 29102-03   Oxyfuel Cutting

Module 29103-03   Base Metal Preparation

Module 29104-03   Weld Quality

Module 29105-03   SMAW – Equipment and Setup

Module 29106-03   SMAW – Electrodes and Selection

Module 29107-03   SMAW – Beads and Fillet Welds

Module 29108-03   SMAW – Groove Welds with Backing

Module 29109-03   Joint Fit-Up and Alignment

Module 29110-03   SMAW – Open V-Groove Welds

Module 29111-03   SMAW – Open-Root Pipe Welds

Note: Modules 29102-03, 29103-03, 29107-03, 29108-03, 29110-03, and 29111-03 include Performance Accreditation Tasks.

Please Address Your Feedback To:

NCCER, Craft Training Department • P.O. Box 141104 • Gainesville, FL 32614-1104
Fax: 352-334-0932   Email: curriculum@nccer.org   Voice: 352-334-0911

Welding Safety

29101-03

Trainee Name: _____

Social Security Number: _____ **Date:** _____

_____ 1. The standard for safety in welding is _____.

    a. *Section II of ASME Boiler and Pressure Vessel Code*

    b. *ANSI Z49.1-1999, Safety in Welding, Cutting, and Allied Processes*

    c. *American Welding Society Structural Welding Code D1.1*

    d. *API 1104, Standard for Welding Pipelines and Related Facilities*

_____ 2. Foot and vehicle traffic in the workplace is _____.

    a. not usually a factor when determining hazards

    b. of particular concern for workers over eighteen

    c. more hazardous at starting and quitting times

    d. regulated by the AHJ

_____ 3. All of the following fabrics may be worn around sources of sparks or high heat *except* _____.

    a. cotton

    b. wool

    c. polyester

    d. leather

_____ 4. Compared to a full leather welding jacket, a welding cape is _____.

    a. cooler

    b. warmer

    c. more protective

    d. more durable

_____ 5. The most effective hearing protection is provided by _____.

    a. earmuff-type hearing protectors

    b. rolled-up cotton balls

    c. reusable earplugs

    d. disposable earplugs

_____ 6. The risk of flash burn can be reduced by wearing _____.

    a. long sleeves and buttoning the shirt all the way to the collar

    b. natural materials like wool or cotton

    c. a welding cap with the visor turned back

    d. safety goggles with a properly tinted lens

_____ 7. An exhaust hood should be located _____ the source of welding fumes.

    a. as close as possible to

    b. four to six inches away from

    c. three to four feet away from

    d. anywhere above

_____　8.　A charcoal filter in an air-purifying respirator should be changed _____.

　　a.　after every eight-hour work shift
　　b.　after 24 hours of total use
　　c.　when the wearer detects any taste or smell
　　d.　three days after the filter cartridge is installed

_____　9.　The proper selection of a respirator is based on the _____.

　　a.　user's size and level of activity
　　b.　contaminant present and its concentration
　　c.　amount of time it will be worn
　　d.　wearer's personal preference

_____　10.　An oxygen level of _____ by volume is considered normal and safe to work in.

　　a.　< 6%
　　b.　6% to 16%
　　c.　19.5% to 23.5%
　　d.　> 23.5%

_____　11.　In addition to protecting other workers from exposure to electric arcs, welding screens _____.

　　a.　prevent drafts from interfering with the stability of the arc
　　b.　reduce glare on the workpiece
　　c.　identify the limits of a welder's work area
　　d.　support workpieces for overhead welding

_____　12.　A _____ is a document issued by the site manager that allows a welder to perform operations with a potential risk of fire.

　　a.　welder qualification record
　　b.　confined space entry permit
　　c.　welding procedure specification
　　d.　hot work permit

_____　13.　Most welding environment fires occur during _____.

　　a.　gas tungsten arc welding
　　b.　oxyfuel gas welding or cutting
　　c.　shielded metal arc welding
　　d.　gas metal arc welding

_____　14.　The storage area for oxygen cylinders must be 20 feet away from fuel gas cylinders or be _____.

　　a.　enclosed on three sides by reinforced concrete walls
　　b.　an enclosed space with a separate ventilation source
　　c.　equipped with an automatic fire suppression system
　　d.　separated by a wall five feet high with a half-hour burn rating

_____　　15.　If acetylene is withdrawn from a cylinder at a rate of more than one-tenth of the volume per hour, there is a risk of _____.

　　a.　dropping the cylinder pressure below 15 psi, causing it to explode
　　b.　frost forming on the regulator and rendering it ineffective
　　c.　the acetylene igniting as it passes through the regulator
　　d.　liquid acetone being drawn into the hoses and gumming up the check valves and regulator

_____　　16.　The outcome of an electric shock is determined by the _____.

　　a.　length of time that electricity passes through the body
　　b.　amount of voltage that passes through the body
　　c.　amount of current that passes through the body
　　d.　polarity of the power source: DCEP, DCEN, or AC

_____　　17.　Detailed information regarding the possible hazards associated with a product is contained in a _____.

　　a.　material safety data sheet (MSDS)
　　b.　confined space entry permit (CSEP)
　　c.　welding procedure specification (WPS)
　　d.　product qualification record (PQR)

_____　　18.　The best method to use when lifting heavy objects is to _____.

　　a.　bend at the waist, grasp the object, and stand up using your back muscles to do the lifting
　　b.　bend at the waist, grasp the object firmly, lift it to your chest and carry it without straightening your back
　　c.　squat down, grasp the object firmly, and stand up by straightening your legs while keeping your back straight
　　d.　kneel down, use your arms to lift the object to your shoulder, and then stand up using your free hand for balance

_____ 19. When using a crane to move large welded assemblies, _____.

    a. pull the load whenever possible

    b. push the load whenever possible

    c. pull the load with your hands if it cannot be pushed

    d. stand behind the load to stabilize it

_____ 20. The first thing to be done at any jobsite before work begins is to _____.

    a. organize the tools and equipment to be used

    b. request a hot work permit

    c. request a confined space entry permit

    d. conduct an evaluation of the hazards that exist

notes

ANSWER KEY

<u>Answer</u>	<u>Section Reference</u>
1. b	1.0.0
2. c	2.2.0
3. c	3.1.0
4. a	3.1.0
5. a	3.4.0
6. d	3.5.0
7. b	4.1.0
8. c	4.2.1
9. b	4.3.0
10. c	5.0.0
11. a	6.0.0
12. d	7.0.0
13. b	8.0.0
14. d	10.0.0
15. d	10.4.0
16. c	12.0.0
17. a	13.0.0
18. c	14.1.0
19. b	14.2.0
20. d	15.0.0

Oxyfuel Cutting

29102-03

Trainee Name: _____

Social Security Number: _____ **Date:** _____

_____ 1. When welding, wear high-top safety boots at least _____ inches tall.

    a. four
    b. five
    c. six
    d. eight

_____ 2. For long-term cutting of materials containing cadmium or mercury, always wear an approved _____.

    a. supplied-air respirator
    b. metal-fume filter
    c. HEPA-rated filter
    d. standard respirator

_____ 3. Before cutting or welding containers, purge them with a flow of _____.

    a. oxygen
    b. fuel gas
    c. inert gas
    d. acetone

_____ 4. Combining pure oxygen with a burning material causes a fire to _____.

    a. flare and burn out of control
    b. extinguish
    c. die down and then flare up
    d. burn more slowly

_____ 5. Oxygen cylinders must be tested _____.

    a. annually
    b. every three years
    c. every five years
    d. every 10 years

_____ 6. The largest oxygen cylinder holds about _____ cubic feet of oxygen.

    a. 85
    b. 285
    c. 400
    d. 485

_____ 7. The most common size oxygen cylinder used for welding and cutting operations holds _____ cubic feet of oxygen.

    a. 85
    b. 227
    c. 485
    d. 2,000

_____ 8. The name of the gas contained inside an oxygen cylinder is stamped, labeled, or stenciled on the _____ of each cylinder.

    a. top
    b. side
    c. bottom
    d. shoulder

_____ 9. When acetylene gas is combined with oxygen during oxyfuel cutting activities, the acetylene causes the flame to burn _____.

   a. hotter than with oxygen combined with propane
   b. at a lower temperature than oxygen alone
   c. up to 4,500°F
   d. hotter than 5,500°F

_____ 10. If an acetylene cylinder is tipped over, stand the cylinder upright and wait at least _____ minutes before using it.

   a. 10
   b. 15
   c. 30
   d. 45

_____ 11. In the top and bottom of the acetylene cylinder are safety plugs with a melting point of _____.

   a. 220°F
   b. 260°F
   c. 300°F
   d. 330°F

_____ 12. Methylacetylene propadiene (MAPP®) gas is a combination of acetylene and _____.

   a. natural gas
   b. propylene
   c. propane
   d. oxygen

_____ 13. When used in oxyfuel cutting, propylene mixtures burn at temperatures around _____.

   a. 2,511°F
   b. 3,563°F
   c. 4,580°F
   d. 5,193°F

_____ 14. If a high volume of gas is removed from a liquefied fuel gas cylinder, the pressure _____.

   a. drops and the temperature of the cylinder also drops
   b. drops and the temperature of the cylinder rises
   c. rises and the temperature of the cylinder also rises
   d. rises and the temperature of the cylinder drops

_____ 15. The regulators used on oxygen cylinders are often painted green and always have _____ on all connections.

   a. right-hand threads
   b. left-hand threads
   c. metric threads
   d. safety latches

_____ 16. As a reminder that the regulator on a cylinder has left-handed threads, a V-notch is sometimes cut around the _____.

   a. regulator
   b. cylinder
   c. threads
   d. nut

_____ 17. The hoses used to carry oxygen from the cylinder to the torch are usually _____.

a. red
b. blue
c. green or black
d. orange and black

_____ 18. On a combination cutting torch, the primary fuel gas and oxygen valves are located on the _____.

a. torch handle
b. cutting head
c. cutting attachment
d. cutting oxygen lever

_____ 19. The selection of a cutting torch tip depends on the thickness of the base metal being cut and the _____ being used.

a. size of fuel hoses
b. size of the torch
c. kind of fuel gas
d. size of the oxygen cylinder

_____ 20. The special cutting tip for cutting off bolt heads and nuts is called a _____ tip.

a. gouging
b. flue cutting
c. rivet cutting
d. riser cutting

_____ 21. An oxyfuel cutting torch should be checked for leaks _____.

a. each time it is used
b. once a week
c. every month
d. after 1,000 hours of use

_____ 22. A cutting flame that has an excess of oxygen is called a(n) _____ flame.

a. cold
b. neutral
c. carburizing
d. oxidizing

_____ 23. When a cutting job is completed and the oxyfuel equipment is no longer needed, the first step in shutting down the oxyfuel cutting equipment is to _____.

a. close the fuel gas cylinder valve
b. back out the fuel gas and oxygen regulator adjusting screws
c. open the fuel gas torch valve
d. open the oxygen torch valve

_____ 24. When an empty cylinder has been removed from the workstation and stored, it must be marked *MT* (or the acceptable site notation for indicating an empty cylinder) _____.

a. near its top
b. on the middle of its side
c. on its bottom
d. on its lower third

_____ 25. When cutting thin steel with oxyfuel cutting equipment, hold the torch so that the tip is pointing in the direction the torch is traveling at a _____ degree angle.

a. 10- to 12-
b. 15- to 20-
c. 21- to 25-
d. 30- to 35-

Craft: Welding

Module Number: 29102-03

Module Title: Oxyfuel Cutting

TRAINEE NAME: _____

TRAINEE SOCIAL SECURITY NUMBER: _____

CLASS: _____

TRAINING PROGRAM SPONSOR: _____

_____

INSTRUCTOR: _____

Rating Levels: 1. Passed: performed task.

2. Failed: did not perform task.

Recognition: When testing for the NCCER Standardized Craft Training Program, be sure to record Performance Profile testing results on Craft Training Report Form 200 and submit the results to the Training Program Sponsor.

Objective	TASK	RATING
2	1. Set up oxyfuel equipment.	
3	2. Light and adjust an oxyfuel torch.	
4	3. Shut down oxyfuel cutting equipment.	
5	4. Disassemble oxyfuel equipment.	
6	5. Change empty cylinders.	

continued

Craft: Welding

Module Number: 29102-03

Module Title: Oxyfuel Cutting

Objective	TASK	RATING
7	6. Perform straight line and square shape cutting.	
7	7. Perform piercing and slot cutting.	
7	8. Perform bevel cutting.	
7	9. Perform washing.	
7	10. Perform gouging.	

ANSWER KEY

Answer	Section Reference
1. d	2.1.0
2. a	2.3.0
3. c	2.2.0
4. a	3.1.0
5. d	3.1.1
6. d	3.1.1
7. b	3.1.1
8. d	3.1.1
9. d	3.2.0
10. c	3.2.1
11. a	3.2.1
12. c	3.3.0
13. d	3.3.0
14. a	3.3.1
15. a	3.4.0
16. d	3.4.0
17. c	3.5.0
18. a	3.6.2
19. c	3.7.0
20. c	3.7.3
21. a	4.12.0
22. d	5.1.0
23. a	6.0.0
24. a	8.0.0
25. b	9.3.0

notes

Oxyfuel Cutting

29102-03

Performance Accreditation Tasks

SETTING UP, IGNITING, ADJUSTING, AND SHUTTING DOWN OXYFUEL EQUIPMENT

2.11

Using oxyfuel equipment that has been completely disassembled, demonstrate how to:

- Set up oxyfuel equipment
- Ignite and adjust the flame
  - Carburizing
  - Neutral
  - Oxidizing
- Shut off the torch
- Shut down the oxyfuel equipment

Criteria for Acceptance

- Set up the oxyfuel equipment in the correct sequence.   _____
- Demonstrate that there are no leaks.   _____
- Properly adjust all three flames.   _____
- Shut off the torch in the correct sequence.   _____
- Shut down the oxyfuel equipment.   _____

CUTTING A SHAPE FROM THIN STEEL

Using a carbon steel plate, lay out and cut the shape and holes shown in the figure. If available, use a machine track cutter to straight cut the 6" dimensions.

NOTE: MATERIAL – CARBON STEEL ⅛" TO ¼" THICK
　　　　　HOLES ¾" DIAMETER
　　　　　SLOTS ¾" × 1½"

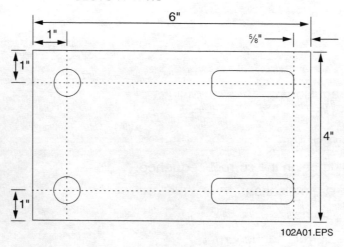

102A01.EPS

Criteria for Acceptance

- Outside dimensions ±1⁄16"
- Inside dimensions (holes and slots) ±⅛"
- Square ±5°
- Minimal amount of slag sticking to plate which can be easily removed
- Square kerf face with minimal notching not exceeding 1⁄16" deep

CUTTING A SHAPE FROM THICK STEEL

Using a carbon steel plate, lay out and cut the shape and holes shown in the figure. If available, use a machine track cutter to bevel and straight cut the 6" dimensions.

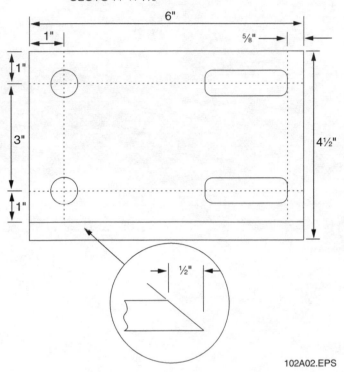

NOTE: MATERIAL – CARBON STEEL OVER ¼" THICK
        HOLES ¾" DIAMETER
        SLOTS ¾" × 1½"

102A02.EPS

Criteria for Acceptance

- Outside dimensions ±⅟₁₆"   _____

- Inside dimensions (holes and slots) ±⅛"   _____

- Square ±5°   _____

- Bevel ± 2°   _____

- Minimal amount of slag sticking to plate, which can be easily removed   _____

- Square kerf face with minimal notching not exceeding ⅟₁₆" deep   _____

Base Metal Preparation

29103-03

Trainee Name: _____

Social Security Number: _____ Date: _____

_____ 1. The most common defect caused by surface contamination is _____.

  a. scale
  b. porosity
  c. brittleness
  d. layering

_____ 2. Wire brushing will not remove _____.

  a. paint
  b. tight corrosion
  c. surface corrosion
  d. light to medium corrosion

_____ 3. Large areas of surface corrosion are best removed by _____.

  a. die grinders
  b. angle grinders and end grinders
  c. needle scalers
  d. weld flux chippers

_____ 4. Surfacing welds are used to _____.

  a. weld stainless steel
  b. make groove welds
  c. build up surfaces
  d. make plug welds

_____ 5. Plug and slot welds are used with _____ joints.

  a. butt
  b. groove
  c. lap
  d. V-groove

_____ 6. Compared to a single V-groove weld, a double V-groove weld _____.

  a. has greater distortion
  b. has equal distortion
  c. requires twice the weld metal
  d. requires half the weld metal

_____ 7. Compared to a fillet weld alone, a combination groove and fillet weld _____.

  a. requires more weld material
  b. saves time and material
  c. requires less preparation time
  d. is more prone to distortion

_____ 8. The groove angle for welding open root welds should be _____ degrees.

  a. 10
  b. 60
  c. 80
  d. 120

_____ 9. The backing strip for welding mild steel up to ¾" thick should be _____.

  a. ¼" thick by ½" wide
  b. ⅜" thick by 3" wide
  c. 1" thick by 5" wide
  d. 2" thick by 3" wide

_____ 10. Always try to position the weldment so that welding is performed in the _____ position.

    a. flat
    b. overhead
    c. vertical
    d. 4G

_____ 11. A detailed listing of the rules and principles that apply to specific welded products is a _____.

    a. welding procedure specification
    b. welding blueprint
    c. welding code
    d. site quality standard

_____ 12. Mechanical joint preparation is _____ than thermal methods of joint preparation.

    a. less precise
    b. slower
    c. faster
    d. more likely to result in oxide deposition

_____ 13. Pipe beveling machines are designed to cut and bevel pipe from _____ in diameter.

    a. ⅞" to 4"
    b. 2" to 5"
    c. 2" to 60"
    d. 3" to 20"

_____ 14. Special internal equipment is available for cutting pipe larger than _____ in diameter.

    a. 18"
    b. 36"
    c. 48"
    d. 54"

_____ 15. Preparing a joint with oxyfuel, plasma arc, or carbon arc cutting processes is known as _____ joint preparation.

    a. automatic
    b. metallurgical
    c. thermal
    d. electrical

contren™
Learning Series

Craft: Welding

Module Number: 29103-03

Module Title: Base Metal Preparation

TRAINEE NAME: _____

TRAINEE SOCIAL SECURITY NUMBER: _____

CLASS: _____

TRAINING PROGRAM SPONSOR: _____

_____

INSTRUCTOR: _____

Rating Levels: 1. Passed: performed task.

2. Failed: did not perform task.

Recognition: When testing for the NCCER Standardized Craft Training Program, be sure to record Performance Profile testing results on Craft Training Report Form 200 and submit the results to the Training Program Sponsor.

Objective	TASK	RATING
4, 5	1. Bevel and prepare the welding coupons for a single open V-groove weld.	

ANSWER KEY

Answer	Section Reference
1. b	2.2.0
2. b	2.3.1
3. b	2.3.2
4. c	3.3.1
5. c	3.3.2
6. d	3.3.7
7. b	3.3.9
8. b	3.3.11
9. b	3.3.12
10. a	3.4.0
11. c	3.5.0
12. b	4.2.0
13. c	4.2.2
14. d	4.3.0
15. c	4.3.0

Base Metal Preparation

29103-03

Performance Accreditation Tasks

PREPARE PLATE JOINTS MECHANICALLY

Using a nibbler or cutter, mechanically prepare the edge of a ¼" to ¾" carbon steel plate with a 22½° or 30° bevel (depending on equipment available) and a ³⁄₃₂" root face as shown.

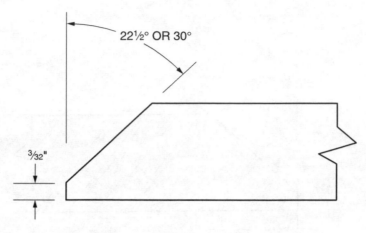

22½° OR 30°

³⁄₃₂"

NOTE: BASE METAL = CARBON STEEL PLATE

103A01.EPS

Criteria for Acceptance:

- Bevel angle ± 2½°   _____
- Bevel face smooth and uniform to ¹⁄₁₆"   _____
- Root face ± ¹⁄₃₂"   _____

PREPARE PIPE JOINTS MECHANICALLY

Using a nibbler or cutter, mechanically prepare the end of a pipe with a 30° or 37½° bevel (depending on the equipment available) and a ³⁄₃₂" root face. Use 6", 8", or 10" Schedule 40 or Schedule 80 carbon steel pipe. The figure shows how to prepare the pipe in this manner.

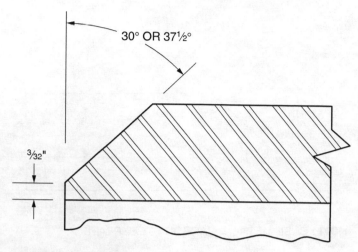

30° OR 37½°

³⁄₃₂"

NOTE: BASE METAL = CARBON STEEL PIPE

103A02.EPS

Criteria for Acceptance:

- Bevel angle ± 2½°   _____

- Bevel face smooth and uniform to ¹⁄₁₆"   _____

- Root face ± ¹⁄₃₂"   _____

PREPARE PLATE JOINTS THERMALLY

Using oxyfuel cutting equipment and a motorized carriage, cut ¼" to ¾" thick carbon steel plate into 3" strips at least 8" long with a bevel of 22½° as shown. Grind a ³⁄₃₂" root face after cutting.

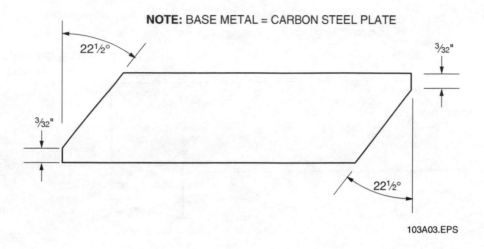

NOTE: BASE METAL = CARBON STEEL PLATE

22½°

³⁄₃₂"

³⁄₃₂"

22½°

103A03.EPS

Criteria for Acceptance:

- Bevel angle ± 2½° _____
- No slag _____
- Minimal notching not exceeding ¹⁄₁₆" deep on the kerf face _____
- Minimum of ¹⁄₁₆" radius at the top edge and bottom edge of the kerf _____
- Root face ± ¹⁄₃₂" _____

PREPARE PIPE JOINTS THERMALLY

Using oxyfuel cutting equipment and a motorized carriage, bevel cut 6", 8", or 10" Schedule 40 or Schedule 80 carbon steel pipe into 4" wide rings. The bevel should be 30° with a ³⁄₃₂" root face prepared with a grinder. Prepare the pipe joint as shown.

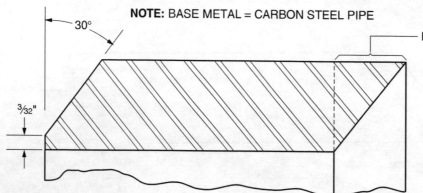

NOTE: BASE METAL = CARBON STEEL PIPE

30°

³⁄₃₂"

NOTE: IF DESIRED, THIS END CAN BE CUT OFF TO SQUARE UP PIPE.

103A04.EPS

Criteria for Acceptance:

- Bevel ± 2½°
- Root face ± ¹⁄₃₂"
- No slag
- Minimal notching not exceeding ¹⁄₁₆" deep on the kerf face
- Minimum of ¹⁄₁₆" radius at the top edge of the kerf

_____

_____

_____

_____

_____

Weld Quality

29104-03

Trainee Name: _____

Social Security Number: _____ **Date:** _____

_____ 1. The specifications for acceptable welding and brazing filler metals and fluxes (Part C) are contained in *Section* _____ of the *American Society of Mechanical Engineers (ASME) Boiler and Pressure Vessel Code (BPVC).*

   a. *I*
   b. *II*
   c. *V*
   d. *IX*

_____ 2. The private organization that does not actually prepare standards, but instead adopts standards that it determines are of value to the public interest is called the _____.

   a. American Society of Mechanical Engineers (ASME)
   b. American National Standards Institute (ANSI)
   c. American Petroleum Institute (API)
   d. American Welding Society (AWS)

_____ 3. Essential variables identified by a welding procedure specification (WPS) are items that cannot be changed without _____.

   a. supervisor approval
   b. site manager approval
   c. requalifying the welder
   d. requalifying the welding procedure

_____ 4. An example of a nonessential variable identified by a WPS is _____.

   a. joint design
   b. travel speed
   c. welding process
   d. material thickness

_____ 5. The results of testing the WPS for a job are recorded on a _____ and kept on file.

   a. work permit
   b. nondestructive testing (NDT) record
   c. procedure qualification record (PQR)
   d. material safety data sheet (MSDS)

_____ 6. The presence of voids or empty spots in the weld metal is called _____.

   a. inclusions
   b. arc strikes
   c. undercut
   d. porosity

_____ 7. Porosity can be grouped into _____ major types.

   a. two
   b. four
   c. six
   d. seven

_____ 8. Most porosity is caused by improper welding techniques and _____.

    a. shielding gas
    b. cold
    c. metal fatigue
    d. contamination

_____ 9. The failure of the filler metal to penetrate and fuse with an area of the weld joint is known as _____.

    a. insufficient heat at the root of the joint
    b. incomplete joint penetration
    c. incomplete fusion
    d. contamination

_____ 10. A groove is melted into the base metal beside the weld in the condition called _____.

    a. cracking
    b. undercut
    c. arc strike
    d. incomplete fusion

_____ 11. Proper visual inspections can detect _____ % of the discontinuities of a weld.

    a. less than 25
    b. up to 50
    c. over 75
    d. 100

_____ 12. The dimensional accuracy of a weld is first checked using _____ examination.

    a. liquid penetrant
    b. radiographic
    c. ultrasonic
    d. visual

_____ 13. A complete set of fillet weld gauges includes seven individual blade gauges that can measure _____ different weld sizes.

    a. 5
    b. 7
    c. 9
    d. 11

_____ 14. The number/letter designator for fillet welding on plate in the flat position is _____.

    a. 3F
    b. 1G
    c. 3G
    d. 1F

_____ 15. In a typical ASME welder qualification test using an open root, if F3 electrodes are used all the way out, the welder qualifies on _____ on both plate and pipe.

    a. F3 electrodes only
    b. F4 electrodes
    c. F3 and F4 electrodes
    d. all electrodes

notes

ANSWER KEY

Answer	Section Reference
1. b	2.1.0
2. b	2.4.0
3. d	3.1.0
4. b	3.1.0
5. c	3.1.0
6. d	4.1.0
7. b	4.1.0
8. d	4.1.0
9. b	4.4.0
10. b	4.6.0
11. c	5.1.0
12. d	5.1.0
13. d	5.1.3
14. d	7.1.0
15. a	7.3.0

SMAW – Equipment and Setup

29105-03

Trainee Name: _____

Social Security Number: _____　　**Date:** _____

_____　1.　If the SMAW electrode tip is held against the workpiece, _____.

a.　the electrode will be heated to a molten state
b.　no welding will take place
c.　an arc will be created
d.　no current will pass through the circuit

_____　2.　There are/is _____ type(s) of welding current.

a.　one
b.　two
c.　three
d.　four

_____　3.　In the United States, AC current is almost always _____ cycles per second.

a.　25
b.　40
c.　50
d.　60

_____　4.　A rectifier converts _____.

a.　high voltage to low voltage
b.　alternating current to direct current
c.　DCEN to DCEP
d.　open circuit voltage to operating voltage

_____　5.　Direct current electrode positive polarity is recommended for _____.

a.　less penetration
b.　flat position welding
c.　out-of-position welding
d.　most types of welding

_____　6.　The primary current for a welding machine can be 120- or 240-volt single-phase or _____-volt three-phase.

a.　40
b.　80
c.　360
d.　480

_____　7.　A step-down transformer converts _____.

a.　high-voltage, high-amperage current to low-voltage, low-amperage current
b.　low-voltage, high-amperage current to high-voltage, low-amperage current
c.　low-voltage, low-amperage current to high-voltage, high-amperage current
d.　high-voltage, low-amperage current to low-voltage, high-amperage current

8. The size of an arc welding machine is determined by its rated amperage and _____.

    a. duty cycle
    b. voltage
    c. electrode size
    d. function

9. The conductors in welding cables are made of strands of _____ wire.

    a. steel
    b. aluminum
    c. copper
    d. stainless steel

10. The longer the distance the welding current must travel, the _____.

    a. larger the welding cable must be to reduce voltage drop
    b. larger the welding cable must be to increase voltage drop
    c. smaller the welding cable must be to reduce voltage drop
    d. smaller the welding cable must be to increase voltage drop

11. Engine generators should be shut down if no welding is to take place for more than _____ minutes.

    a. 10
    b. 15
    c. 20
    d. 30

12. The preventive maintenance schedule for an engine-driven welding machine is usually based on _____.

    a. insurance figures
    b. the number of welds made
    c. fuel consumption figures
    d. the number of hours that the engine operates

13. Hand tools for preparing a welding surface and cleaning a completed weld include _____.

    a. files, brushes, and chipping hammers
    b. nibblers
    c. angle grinders
    d. die grinders

14. Weld slag chippers and needle scalers are _____ tools.

    a. pneumatically powered
    b. hand
    c. ultrasonic
    d. battery-powered

15. Weld slag chippers and needle scalers are not very effective at _____.

    a. removing surface corrosion
    b. cleaning surfaces prior to welding
    c. removing hardened dirt from weld surfaces
    d. removing slag from weld

Craft: Welding

Module Number: 29105-03

Module Title: SMAW – Equipment and Setup

contren™ Learning Series

TRAINEE NAME: _____

TRAINEE SOCIAL SECURITY NUMBER: _____

CLASS: _____

TRAINING PROGRAM SPONSOR: _____

_____

INSTRUCTOR: _____

Rating Levels: 1. Passed: performed task.

2. Failed: did not perform task.

Recognition: When testing for the NCCER Standardized Craft Training Program, be sure to record Performance Profile testing results on Craft Training Report Form 200 and submit the results to the Training Program Sponsor.

Objective	TASK	RATING
5	1. Set up a machine for welding.	

ANSWER KEY

<u>Answer</u>	<u>Section Reference</u>
1. b	2.0.0
2. b	3.0.0
3. d	3.1.0
4. b	3.2.0
5. c	3.2.1
6. d	6.0.0
7. d	6.1.0
8. a	7.0.0
9. c	8.0.0
10. a	8.0.0
11. d	9.7.3
12. d	9.7.4
13. a	10.1.0
14. a	10.2.0
15. a	10.2.0

SMAW – Electrodes and Selection

29106-03

Trainee Name: _____

Social Security Number: _____   Date: _____

_____   1. The majority of the weld is formed by the _____.

   a. wire core
   b. base metal mixing with the flux
   c. flux coating
   d. wire core mixing with the melted base metal

_____   2. The flux coating of electrodes _____ the molten metal.

   a. cools
   b. smooths
   c. cleans and oxidizes
   d. cleans and deoxidizes

_____   3. The powdered metal added to some flux coatings adds filler metal to a weld and helps _____ the weld.

   a. improve the appearance of
   b. reduce spatter at
   c. clean
   d. cool

_____   4. The most common electrodes are discussed in the American Welding Society (AWS) _____ classifications.

   a. A5.1 and A5.3
   b. A5.1 and A5.5
   c. A5.11
   d. A5.24

_____   5. To determine the minimum tensile strength per square inch of the deposited weld metal created by an electrode, multiply the first two digits (or three if 100 or over) of the marking on the electrode by _____.

   a. the last two digits
   b. the last digit
   c. 100
   d. 1,000

_____   6. All electrodes manufactured in the United States must have the AWS classification number printed on them as well as on the _____.

   a. container in which they are shipped
   b. shipping documents
   c. electrode holder
   d. welding permit

_____   7. Electrodes with powdered iron in their flux belong to the _____ electrode group.

   a. fast-fill
   b. fill-freeze
   c. fast-freeze
   d. low-hydrogen

_____ 8. The chemical and mechanical properties of the base metal can be obtained by referring to the _____.

  a. welding procedure specification (WPS)
  b. base metal mill specifications
  c. manufacturers' classification
  d. welding permit

_____ 9. Normally an electrode should be _____than the metal being welded.

  a. higher in tensile strength
  b. smaller in diameter
  c. larger in diameter
  d. equal in diameter

_____ 10. After a sealed container of low-hydrodes electrodes is opened, the electrodes must be stored in special ovens at a temperature setting of _____.

  a. 100°F
  b. 150°F
  c. 200°F
  d. 250°F

_____ 11. Non-low-hydrogen electrodes may be used if their coatings are not damaged and they have not come in direct contact with _____.

  a. the ground
  b. humidity
  c. water or oil
  d. direct sunlight

_____ 12. The period of time a low-hydrogen electrode can be outside a heated holding or portable oven is _____.

  a. limited to one hour
  b. limited to two hours
  c. defined as storage time
  d. defined as exposure time

_____ 13. Low-hydrogen electrodes that exceed the maximum exposure time can be _____.

  a. used anyway
  b. dried one time
  c. dried repeatedly
  d. used for jobs requiring lower standards

_____ 14. Electrodes not subject to X-ray inspections must be brought up to _____ and held at that temperature for at least one hour.

  a. 250°F
  b. 300°F
  c. 400°F
  d. 750°F

_____ 15. Acceptable base metal and filler metal combinations are listed in _Section II, Material Specifications_ of the _____ .

  a. OSHA Requirements on Electrical Grounding
  b. _American Society of Mechanical Engineers (ASME) Boiler and Pressure Vessel Code_
  c. _Lincoln Rod Book_
  d. _Procedure Handbook of Arc Welding_

notes

Answer	**Section Reference**
1. d	2.0.0
2. d	2.0.0
3. a	2.0.0
4. b	3.0.0
5. d	3.1.0
6. a	3.2.0
7. a	4.1.2
8. b	4.2.2
9. b	4.2.3
10. d	5.1.0
11. c	5.2.0
12. d	5.4.1
13. b	5.4.2
14. b	5.4.2
15. b	6.0.0

SMAW – Beads
and Fillet Welds

29107-03

Trainee Name: _____

Social Security Number: _____　　**Date:** _____

_____　　1.　Before welding coupons made of ¼" to ¾" mild steel, use a wire brush or grinder to _____ the coupons.

　　a.　shape the edges of
　　b.　polish the surface of
　　c.　rough the surface of
　　d.　remove heavy mill scale or corrosion from

_____　　2.　When practicing welding, reuse weld coupons until _____.

　　a.　all surfaces have been welded on
　　b.　they cannot be cut apart
　　c.　they are needed for a job
　　d.　they are burned

_____　　3.　When practicing welding with an AC welding machine, use _____ electrodes.

　　a.　⅛" E7018
　　b.　⁵⁄₃₂" E7018
　　c.　E6011 and E7018
　　d.　E6011 and E6013

_____　　4.　Arc length is measured from _____.

　　a.　the end of the electrode holder to the base metal
　　b.　the end of the electrode core to the base metal
　　c.　one end of the electrode to the other
　　d.　the striking area to the ground clamp

_____　　5.　The best method of striking an arc when using a transformer DC welding machine is the _____ method.

　　a.　scratching
　　b.　grounding
　　c.　weaving
　　d.　tapping

_____　　6.　Arc strikes are left on the base metal by the _____ method of striking an arc.

　　a.　scratching
　　b.　grounding
　　c.　weaving
　　d.　tapping

_____　　7.　Concentrated magnetic fields in the base metal cause the arc to _____.

　　a.　stop
　　b.　increase
　　c.　decrease
　　d.　deflect

_____　　8.　Magnetic fields created during welding tend to concentrate in the corners of the _____.

　　a.　electrode
　　b.　base metal
　　c.　electrode holder
　　d.　grounding clamp

_____ 9. A weld bead that is made with a side-to-side motion of the electrode is called a(n) _____ bead.

    a. arc
    b. weave
    c. stringer
    d. whipping

_____ 10. To control the weld puddle when depositing a stringer bead, use a _____ motion.

    a. short
    b. jerky
    c. constant
    d. whipping

_____ 11. Welding codes require that the crater made by a termination of the weld be filled to the full _____ of the weld.

    a. cross section
    b. length
    c. width
    d. depth

_____ 12. When overlapping stringer beads, use a work angle of _____ degrees toward the first stringer bead to obtain proper tie-in.

    a. 5 to 10
    b. 10 to 15
    c. 15 to 20
    d. 20 to 22

_____ 13. The point at which the back of the weld intersects the base metal is called the weld _____.

    a. leg
    b. toe
    c. face
    d. root

_____ 14. A fillet weld is acceptable even if it has _____.

    a. excessive undercut
    b. a slightly non-uniform face
    c. insufficient leg
    d. incomplete fusion

_____ 15. The most common fillet welds made are in _____ joints.

    a. lap and T-
    b. horizontal
    c. overhead
    d. vertical

contren™
Learning Series

Craft: Welding

Module Number: 29107-03

Module Title: SMAW – Beads and Fillet Welds

TRAINEE NAME: _____

TRAINEE SOCIAL SECURITY NUMBER: _____

CLASS: _____

TRAINING PROGRAM SPONSOR: _____

_____

INSTRUCTOR: _____

Rating Levels: 1. Passed: performed task.

2. Failed: did not perform task.

Recognition: When testing for the NCCER Standardized Craft Training Program, be sure to record Performance Profile testing results on Craft Training Report Form 200 and submit the results to the Training Program Sponsor.

Objective	TASK	RATING
1	1. Set up AC welding equipment.	
2, 3	2. Strike an arc.	
5	3. Make stringer, weave, and overlapping beads using E6010 and E7018 electrodes.	
6	4. Make fillet welds in the 2F, 3F, and 4F positions using E6010 and E7018 electrodes.	

ANSWER KEY

<u>Answer</u>	<u>Section Reference</u>
1. d	3.2.0
2. a	3.2.0
3. d	3.4.0
4. b	4.0.0
5. d	4.2.0
6. a	4.2.0
7. d	5.0.0
8. b	5.0.0
9. b	6.0.0
10. d	6.1.0
11. a	6.4.0
12. b	7.1.0
13. d	8.0.0
14. b	8.0.0
15. a	8.1.0

SMAW– Beads and Fillet Welds

29107-03

Performance Accreditation Tasks

BUILD A PAD WITH E6010 ELECTRODES IN THE FLAT POSITION

Using ⅛" E6010 electrodes, build up a pad of weld metal on carbon steel plate as indicated.

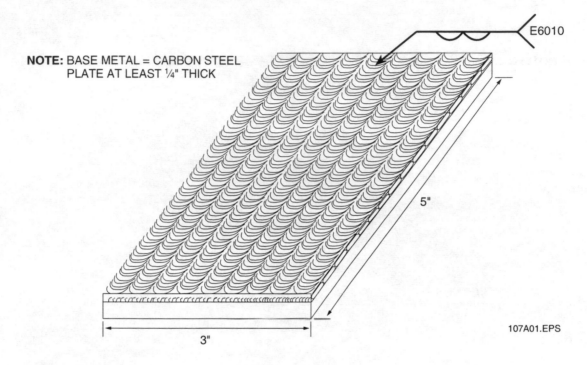

NOTE: BASE METAL = CARBON STEEL PLATE AT LEAST ¼" THICK

E6010

5"

3"

107A01.EPS

Criteria for Acceptance:

- Weld beads straight to within ⅛" _____
- Uniform rippled appearance on the bead face _____
- Craters and restarts filled to the full cross section of the weld _____
- Face of the pad flat to within ⅛" _____
- Smooth flat transition with complete fusion at the toes of one bead into the face of the previous bead _____
- No porosity _____
- No overlap at weld toes _____
- No undercut _____
- No inclusions _____
- No cracks _____

BUILD A PAD WITH E7018 ELECTRODES IN THE FLAT POSITION

Using ⅛" E7018 electrodes, build up a pad of weld metal on carbon steel plate as indicated.

NOTE: BASE METAL = CARBON STEEL PLATE AT LEAST ¼" THICK

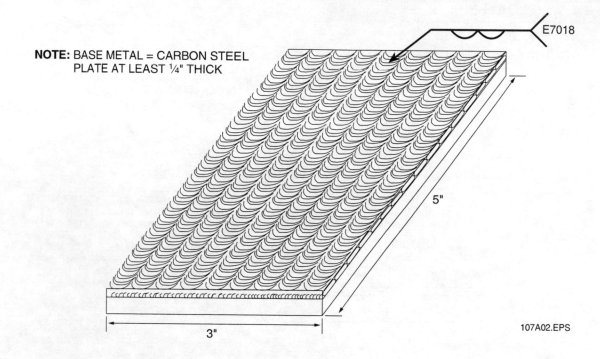

E7018

5"

3"

107A02.EPS

Criteria for Acceptance:

- Weld beads straight to within ⅛"
- Uniform rippled appearance on the bead face
- Craters and restarts filled to the full cross section of the weld
- Face of the pad flat to within ⅛"
- Smooth flat transition with complete fusion at the toes of one bead into the face of the previous bead
- No porosity
- No overlap at weld toes
- No undercut
- No inclusions
- No cracks

HORIZONTAL (2F) FILLET WELD
WITH E6010 ELECTRODES

Using ⅛" E6010 electrodes, make a horizontal fillet weld as indicated.

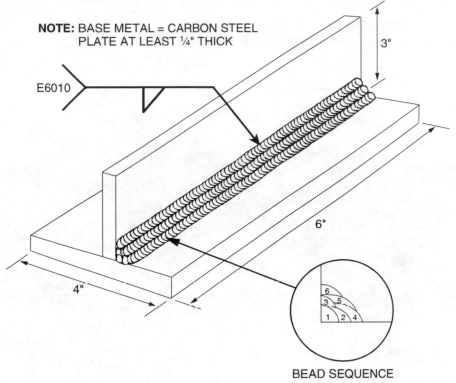

NOTE: BASE METAL = CARBON STEEL
PLATE AT LEAST ¼" THICK

E6010

3"

6"

4"

BEAD SEQUENCE

107A03.EPS

Criteria for Acceptance:

- Uniform rippled appearance on the bead face

- Craters and restarts filled to the full cross section of the weld

- Uniform weld size, ±¹⁄₁₆"

- Acceptable weld profile in accordance with *AWS D1.1*

- Smooth transition with complete fusion at the toes of the weld

- No porosity

- No undercut

- No inclusions

- No cracks

- No overlap

HORIZONTAL (2F) FILLET WELD WITH E7018 ELECTRODES

Using ⅛" E7018 electrodes, make a horizontal fillet weld as indicated.

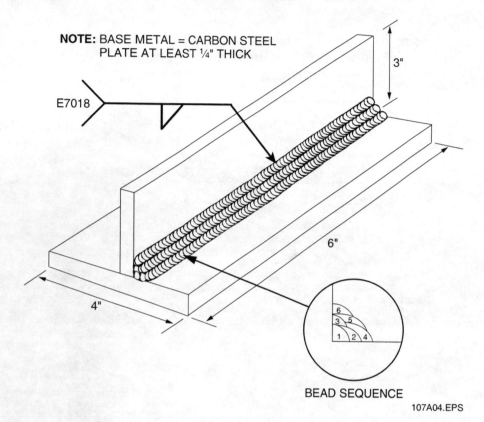

NOTE: BASE METAL = CARBON STEEL PLATE AT LEAST ¼" THICK

E7018

3"

6"

4"

BEAD SEQUENCE

107A04.EPS

Criteria for Acceptance:

- Uniform rippled appearance on the bead face
- Craters and restarts filled to the full cross section of the weld
- Uniform weld size, ±¹⁄₁₆"
- Acceptable weld profile in accordance with *AWS D1.1*
- Smooth transition with complete fusion at the toes of the weld
- No porosity
- No undercut
- No inclusions
- No cracks
- No overlap

_____

_____

_____

_____

_____

_____

_____

_____

_____

_____

VERTICAL (3F) FILLET WELD WITH E6010 ELECTRODES

Using ⅛" E6010 electrodes, make a vertical fillet weld as indicated.

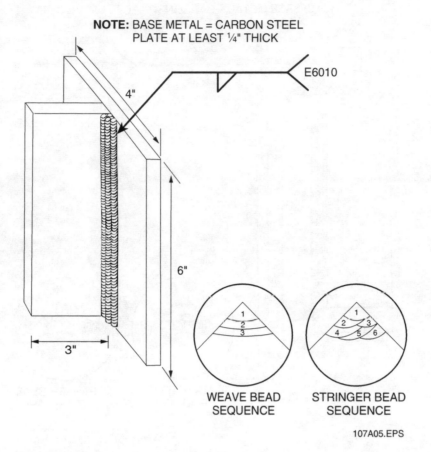

NOTE: BASE METAL = CARBON STEEL PLATE AT LEAST ¼" THICK

E6010

4"

6"

3"

WEAVE BEAD SEQUENCE

STRINGER BEAD SEQUENCE

107A05.EPS

Criteria for Acceptance:

- Uniform rippled appearance on the bead face
- Craters and restarts filled to the full cross section of the weld
- Uniform weld size, ±¹⁄₁₆"
- Acceptable weld profile in accordance with *AWS D1.1*
- Smooth transition with complete fusion at the toes of the weld
- No porosity
- No undercut
- No inclusions
- No cracks
- No overlap at weld toes

VERTICAL (3F) FILLET WELD WITH E7018 ELECTRODES

Using ⅛" E7018 electrodes, make a vertical fillet weld as indicated.

NOTE: BASE METAL = CARBON STEEL PLATE AT LEAST ¼" THICK

WEAVE BEAD SEQUENCE

STRINGER BEAD SEQUENCE

107A06.EPS

Criteria for Acceptance:

- Uniform rippled appearance on the bead face
- Craters and restarts filled to the full cross section of the weld
- Uniform weld size, ±¹⁄₁₆"
- Acceptable weld profile in accordance with *AWS D1.1*
- Smooth transition with complete fusion at the toes of the weld
- No porosity
- No undercut
- No inclusions
- No cracks
- No overlap at weld toes

OVERHEAD (4F) FILLET WELD WITH E6010 ELECTRODES

Using ⅛" E6010 electrodes, make an overhead fillet weld as indicated.

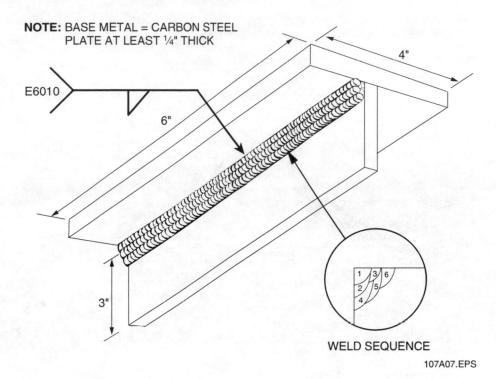

NOTE: BASE METAL = CARBON STEEL PLATE AT LEAST ¼" THICK

E6010

6"

4"

3"

WELD SEQUENCE

107A07.EPS

Criteria for Acceptance:

- Uniform rippled appearance on the bead face   _____
- Craters and restarts filled to the full cross section of the weld   _____
- Uniform weld size, ±1⁄16"   _____
- Acceptable weld profile in accordance with *AWS D1.1*   _____
- Smooth transition with complete fusion at the toes of the weld   _____
- No porosity   _____
- No undercut   _____
- No inclusions   _____
- No cracks   _____
- No overlap   _____

OVERHEAD (4F) FILLET WELD WITH E7018 ELECTRODES

Using ⅛" E7018 electrodes, make an overhead fillet weld as indicated.

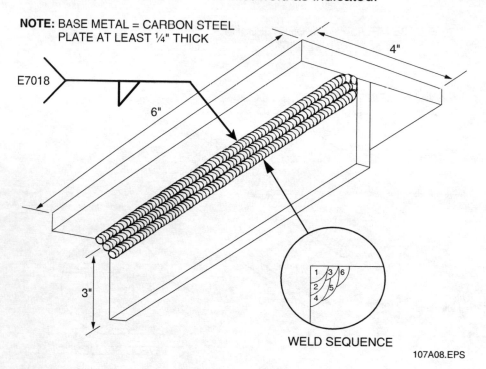

NOTE: BASE METAL = CARBON STEEL PLATE AT LEAST ¼" THICK

E7018

4"

6"

3"

WELD SEQUENCE

107A08.EPS

Criteria for Acceptance:

- Uniform rippled appearance on the bead face
- Craters and restarts filled to the full cross section of the weld
- Uniform weld size, ±¹⁄₁₆"
- Acceptable weld profile in accordance with *AWS D1.1*
- Smooth transition with complete fusion at the toes of the weld
- No porosity
- No undercut
- No inclusions
- No cracks
- No overlap

SMAW – Groove Welds with Backing

29108-03

Trainee Name: _____

Social Security Number: _____   **Date:** _____

_____ 1. Prior to welding, any surface in a weld groove is called a _____.

a. bevel angle
b. groove face
c. bevel radius
d. groove angle

_____ 2. The distance the joint preparation extends into the base metal is called the _____.

a. root of the joint
b. depth of bevel
c. groove radius
d. root opening

_____ 3. Groove welds with metal backing are used for _____ welds.

a. high-strength
b. double groove
c. temporary
d. single pass fillet

_____ 4. To conform with AWS limited-thickness test coupon requirements, weld coupons should be carbon steel that is _____ thick.

a. ¼"
b. ⅜"
c. ½"
d. ¾"

_____ 5. When preparing weld coupons for V-groove welds with backing, cut the metal backing strips so that they are _____.

a. 2" by 8"
b. 3" by 8"
c. 1" by 8"
d. 1" by 3"

_____ 6. When preparing weld coupons for V-groove welds with backing, center the beveled strips on the backing strip with a _____ gap and tack-weld them in place.

a. ⅛"
b. ¼"
c. ⅓"
d. ½"

_____ 7. The most common American Welding Society (AWS) plate welding qualification test requires the use of _____ electrodes.

a. fast-fill
b. fill-freeze
c. fast-freeze
d. low-hydrogen

_____ 8. In the 1G and 2G groove weld positions, the weld axis can be inclined up to _____.

a. 15 degrees
b. 30 degrees
c. 45 degrees
d. 55 degrees

9. Groove welds should be made with slight reinforcement and a gradual transition to the base metal at each _____.

    a. leg
    b. toe
    c. face
    d. root

10. When using weave beads on flat V-groove welds, keep the electrode angle at a 0 degree work angle with a _____ degree drag angle.

    a. 5- to 10-
    b. 10- to 15-
    c. 15- to 20-
    d. 20- to 22-

11. Cooling welds with water can cause weld cracks and affect the mechanical properties of the _____.

    a. pad
    b. root
    c. electrode
    d. base metal

12. When using stringer beads on V-groove welds in the horizontal position, keep the electrode drag angle at _____ degrees.

    a. 10 to 15
    b. 15 to 30
    c. 35 to 40
    d. 45 to 55

13. When practicing horizontal V-groove welds, pay particular attention at the termination of the weld to fill the _____.

    a. pad
    b. root
    c. crater
    d. puddle

14. When running vertical stringer beads on a V-groove joint in the vertical position, keep the electrode at a _____ degree push angle.

    a. 0- to 10-
    b. 5- to 15-
    c. 15- to 20-
    d. 20- to 22-

15. When running vertical stringer beads on a V-groove joint in the vertical position, set the welder amperage in the _____ of the range for better puddle control.

    a. top 10 percent
    b. top 20 percent
    c. middle part
    d. lower part

contren™
Learning Series

Craft: Welding

Module Number: 29108-03

Module Title: SMAW – Groove Welds with Backing

TRAINEE NAME: _____

TRAINEE SOCIAL SECURITY NUMBER: _____

CLASS: _____

TRAINING PROGRAM SPONSOR: _____

_____

INSTRUCTOR: _____

Rating Levels: 1. Passed: performed task.

2. Failed: did not perform task.

Recognition: When testing for the NCCER Standardized Craft Training Program, be sure to record Performance Profile testing results on Craft Training Report Form 200 and submit the results to the Training Program Sponsor.

Objective	TASK	RATING
3	1. Set up the arc welding equipment for making groove welds.	
4	2. Make flat welds on V-groove joints using E7018 electrodes.	
4	3. Make horizontal welds on V-groove joints using E7018 electrodes.	
4	4. Make vertical welds on V-groove joints using E7018 electrodes.	
4	5. Make overhead welds on V-groove joints using E7018 electrodes.	

ANSWER KEY

<u>Answer</u>	<u>Section Reference</u>
1. b	2.3.0
2. b	2.3.0
3. a	2.5.0
4. b	3.3.0
5. c	3.3.0
6. b	3.3.0
7. d	4.0.0
8. a	4.1.0
9. b	4.2.0
10. b	5.1.0
11. d	5.1.0
12. a	5.2.1
13. c	5.2.2
14. a	5.3.1
15. d	5.3.2

SMAW – Groove Welds with Backing

29108-03

Performance Accreditation Tasks

V-GROOVE WELDS WITH BACKING IN THE FLAT (1G) POSITION

Using ³⁄₃₂", ⅛", and ⁵⁄₃₂" E7018 electrodes, make a V-groove weld with steel backing on carbon steel plate in the flat position as indicated.

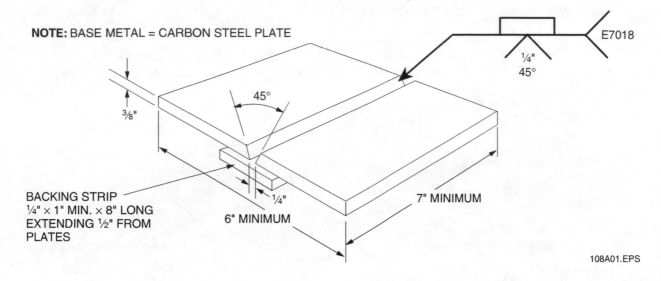

NOTE: BASE METAL = CARBON STEEL PLATE

45°

³⁄₈"

E7018

¼"
45°

BACKING STRIP
¼" × 1" MIN. × 8" LONG
EXTENDING ½" FROM
PLATES

¼"

6" MINIMUM

7" MINIMUM

108A01.EPS

Criteria for Acceptance:

- Uniform rippled appearance on the bead face   _____
- Craters and restarts filled to the full cross section of the weld   _____
- Uniform weld size ±¹⁄₁₆"   _____
- Acceptable weld profile in accordance with *AWS D1.1*   _____
- Smooth transition with complete fusion at the toes of the weld   _____
- No porosity   _____
- No overlap   _____
- No undercut   _____
- No inclusions   _____
- No cracks   _____
- Acceptable guided bend test results   _____

V-GROOVE WELDS WITH BACKING IN THE HORIZONTAL (2G) POSITION

Using ³⁄₃₂", ⅛", and ⁵⁄₃₂" E7018 electrodes, make a V-groove weld with steel backing on carbon steel plate in the horizontal position as indicated.

NOTE: BASE METAL = CARBON STEEL PLATE

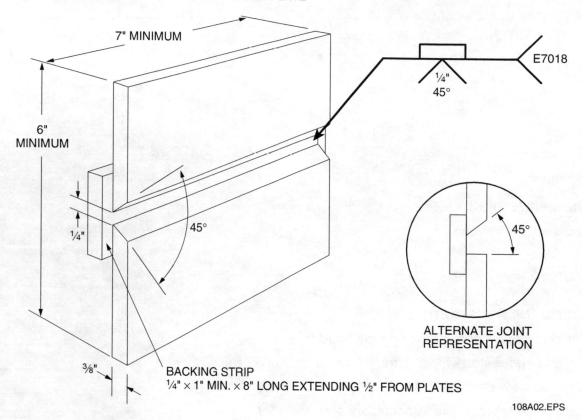

7" MINIMUM

6" MINIMUM

¼"

45°

45°

E7018

¼"

45°

ALTERNATE JOINT REPRESENTATION

⅜"

BACKING STRIP
¼" × 1" MIN. × 8" LONG EXTENDING ½" FROM PLATES

108A02.EPS

Criteria for Acceptance:

- Uniform rippled appearance on the bead face
- Craters and restarts filled to the full cross section of the weld
- Uniform weld size ±¹⁄₁₆"
- Acceptable weld profile in accordance with *AWS D1.1*
- Smooth transition with complete fusion at the toes of the weld
- No porosity
- No undercut
- No overlap
- No inclusions
- No cracks
- Acceptable guided bend test results

V-GROOVE WELD WITH BACKING IN THE VERTICAL (3G) POSITION

Using ³⁄₃₂", ⅛", and ⁵⁄₃₂" E7018 electrodes, make a V-groove weld with steel backing on carbon steel plate in the vertical position as indicated.

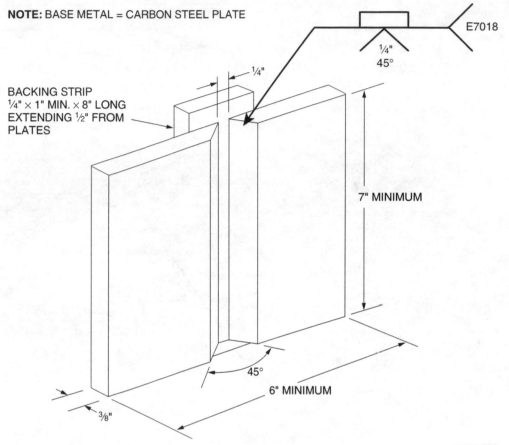

NOTE: BASE METAL = CARBON STEEL PLATE

E7018

¼"
45°

¼"

BACKING STRIP
¼" × 1" MIN. × 8" LONG
EXTENDING ½" FROM
PLATES

7" MINIMUM

45°

6" MINIMUM

³⁄₈"

108A03.EPS

Criteria for Acceptance:

- Uniform rippled appearance on the bead face　　　　　　＿＿＿＿＿＿
- Craters and restarts filled to the full cross section of the weld　　＿＿＿＿＿＿
- Uniform weld size ±¹⁄₁₆"　　　　　　＿＿＿＿＿＿
- Acceptable weld profile in accordance with *AWS D1.1*　　＿＿＿＿＿＿
- Smooth transition with complete fusion at the toes of the weld　　＿＿＿＿＿＿
- No porosity　　　　　　＿＿＿＿＿＿
- No overlap　　　　　　＿＿＿＿＿＿
- No undercut　　　　　　＿＿＿＿＿＿
- No inclusions　　　　　　＿＿＿＿＿＿
- No cracks　　　　　　＿＿＿＿＿＿
- Acceptable guided bend test results　　　　　　＿＿＿＿＿＿

V-GROOVE WELD WITH BACKING IN THE OVERHEAD (4G) POSITION

Using ³⁄₃₂", ⅛", and ⁵⁄₃₂" E7018 electrodes, make a V-groove weld with steel backing on carbon steel plate in the overhead position as indicated.

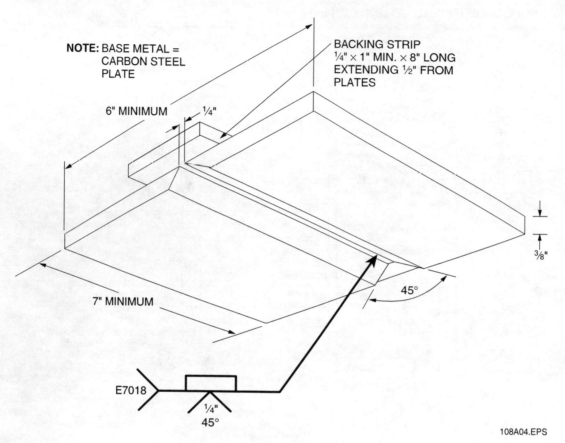

NOTE: BASE METAL = CARBON STEEL PLATE

BACKING STRIP ¼" × 1" MIN. × 8" LONG EXTENDING ½" FROM PLATES

6" MINIMUM   ¼"

7" MINIMUM

45°

³⁄₈"

E7018   ¼"   45°

108A04.EPS

Criteria for Acceptance:

- Uniform rippled appearance on the bead face
- Craters and restarts filled to the full cross section of the weld
- Uniform weld size ±¹⁄₁₆"
- Acceptable weld profile in accordance with *AWS D1.1*
- Smooth transition with complete fusion at the toes of the weld
- No porosity
- No undercut
- No overlap
- No inclusions
- No cracks
- Acceptable guided bend test results

Joint Fit-Up
and Alignment
29109-03

Trainee Name: _____

Social Security Number: _____　**Date:** _____

_____　1.　The letter *A* in front of the year on the cover of the American Society of Mechanical Engineers (ASME) code indicates that the code contains _____.

　　a.　art changes
　　b.　new changes
　　c.　a yearly addendum
　　d.　a semi-annual addendum

_____　2.　Before using a WPS (weld procedure specification), each manufacturer or contractor is required to _____ the WPS.

　　a.　correct
　　b.　copy
　　c.　test and qualify
　　d.　alter as necessary

_____　3.　The results of testing the WPS for a job are recorded on a _____ and kept on file.

　　a.　PQR (procedure qualification record)
　　b.　nondestructive testing (NDT) record
　　c.　job blueprint
　　d.　work permit (or order)

_____　4.　The centering head attachment of a combination square is used to measure _____.

　　a.　round stock and to locate the center of shafts or other round objects
　　b.　90-degree and 45-degree angles
　　c.　joint alignment
　　d.　depth

_____　5.　The tool that uses a bubble in a glass vial to check level and plumb is a _____.

　　a.　level
　　b.　Hi-Lo gauge
　　c.　straightedge
　　d.　combination square

_____　6.　Internal misalignment between two sections of pipe to be welded end to end is checked using a _____.

　　a.　torpedo level
　　b.　combination square
　　c.　protractor
　　d.　Hi-Lo gauge

_____　7.　When using a chain fall to lift or lower weldment parts, always secure the chain fall to an overhead _____.

　　a.　pipe
　　b.　duct
　　c.　conduit
　　d.　structural member

_____ 8. The gap plate of a plate alignment tool can be changed to match the specified _____.

a. root bar
b. joint angle
c. root opening
d. joint thickness

_____ 9. Chain clamps are used to align and hold pipe for _____.

a. bending
b. rotating and cutting
c. cleaning
d. fit-up and tacking

_____ 10. Metals that show relatively greater expansion and contraction when heated are _____ than other metals.

a. weaker
b. stronger
c. less likely to become distorted
d. more likely to become distorted

_____ 11. Field-fabricated support devices are often used during welding to _____.

a. eliminate residual stress in the weldment
b. prevent warping caused by distortion
c. increase the coefficient of thermal expansion
d. control burn-through

_____ 12. Excessive face reinforcement _____.

a. stabilizes the weld
b. is mandated by welding codes
c. reduces the strength of a weld
d. is permitted by codes but wastes weld material

_____ 13. When aligning a joint fit-up, an open root joint should have a root opening from _____.

a. $\frac{1}{32}$" to $\frac{1}{16}$"
b. $\frac{1}{16}$" to $\frac{1}{8}$"
c. $\frac{1}{8}$" to $\frac{1}{4}$"
d. $\frac{1}{4}$" to $\frac{1}{2}$"

_____ 14. A simple welding sequence with short welds _____ of the joint is recommended to reduce the effects of shrinkage.

a. on alternating sides
b. on the ends
c. on one side and long welds on the other side
d. in the middle and long welds on the ends

_____ 15. The procedure in which two individuals perform welding sequences on opposite sides of a joint at the same time is _____.

a. a very effective way to control distortion
b. prohibited by most welding codes
c. known as backstep welding
d. likely to result in distortion

contren™
Learning Series

Craft: Welding

Module Number: 29109-03

Module Title: Joint Fit-Up and Alignment

TRAINEE NAME: _____

TRAINEE SOCIAL SECURITY NUMBER: _____

CLASS: _____

TRAINING PROGRAM SPONSOR: _____

_____

INSTRUCTOR: _____

Rating Levels:	1. Passed: performed task.	
	2. Failed: did not perform task.	
Recognition:	When testing for the NCCER Standardized Craft Training Program, be sure to record Performance Profile testing results on Craft Training Report Form 200 and submit the results to the Training Program Sponsor.	

Objective	TASK	RATING
4	1. Fit up joints using plate and pipe fit-up tools.	
2, 5	2. Check the joint for proper fit-up and alignment using gauges and measuring devices.	

ANSWER KEY

Answer	Section Reference
1. c	2.2.0
2. c	2.3.0
3. a	2.3.0
4. a	3.2.0
5. a	3.3.0
6. d	3.4.0
7. d	4.1.2
8. c	4.2.4
9. d	4.3.2
10. d	5.2.0
11. b	5.3.1
12. c	5.3.3
13. b	5.3.3
14. a	5.3.11
15. a	5.3.11

SMAW – Open V-Groove Welds

29110-03

Trainee Name: _____

Social Security Number: _____   **Date:** _____

_____ 1. If ⅜" carbon steel is not available, practice welds can be made on steel _____ to ¾" thick.

a. ⅛"
b. ¼"
c. ⅓"
d. ½"

_____ 2. Welding codes allow the _____ on open-root V-groove welds to be within the range of 0 to ⅛".

a. leg
b. toe
c. root face
d. throat

_____ 3. For better puddle control when welding vertically, you should _____.

a. set the amperage in the upper part of the suggested range
b. set the amperage in the lower part of the suggested range
c. maintain a long arc
d. use a whipping motion with low-hydrogen electrodes

_____ 4. The most difficult part of making an open-root V-groove weld is the _____.

a. bead
b. overlap
c. keyhole
d. root pass

_____ 5. For an open-root V-groove weld, root reinforcement should be flush to within _____.

a. ¹⁄₃₂"
b. ¹⁄₁₆"
c. ⅛"
d. ³⁄₁₆"

_____ 6. When the plates are vertical and the axis of the weld is horizontal, the code for this position of groove weld is _____.

a. 1G
b. 2G
c. 3G
d. 4G

_____ 7. In the 1G and 2G positions, the weld axis can be inclined up to _____ degrees.

a. 15
b. 25
c. 35
d. 45

_____ 8. Groove welds should be made with slight reinforcement and a gradual transition to the base metal at each _____.

a. weld root
b. throat
c. toe
d. root opening

9. When using weave beads in flat open-root V-groove welds, use a _____ degree drag angle.

   a. 10- to 15-
   b. 15- to 20-
   c. 20- to 30-
   d. 35- to 40-

10. Cooling with water is only done on _____.

    a. test coupons
    b. on-the-job welds
    c. practice coupons
    d. horizontal welds

11. Horizontal welds can be made on a horizontal open-root V-groove or on a(n) _____.

    a. overhead open-root V-groove
    b. overhead flat surface
    c. vertical open-root V-groove
    d. vertically positioned flat surface

12. To control the weld puddle when welding horizontal open-root V-groove welds, drop the angle of the electrode slightly, but no more than _____ degrees.

    a. 5
    b. 10
    c. 15
    d. 20

13. When using weave beads on open-root V-groove welds in the vertical position, keep the electrode at _____ degrees to the plate surface for the first pass.

    a. 30
    b. 45
    c. 60
    d. 90

14. When running vertical weave beads on a V-groove joint in the vertical position, the width of the weave beads should be no more than _____ time(s) the diameter of the electrode being used.

    a. one
    b. three
    c. four
    d. five

15. To prevent undercut in a vertical open-root V-groove weld, the welder should _____.

    a. increase the arc length
    b. use a downward push angle
    c. use a 30-degree drag angle
    d. pause slightly at each weld toe to penetrate and fill the crater

contren™
Learning Series

Craft: Welding

Module Number: 29110-03

Module Title: SMAW – Open V-Groove Welds

TRAINEE NAME: _____

TRAINEE SOCIAL SECURITY NUMBER: _____

CLASS: _____

TRAINING PROGRAM SPONSOR: _____

_____

INSTRUCTOR: _____

Rating Levels: 1. Passed: performed task.

2. Failed: did not perform task.

Recognition: When testing for the NCCER Standardized Craft Training Program, be sure to record Performance Profile testing results on Craft Training Report Form 200 and submit the results to the Training Program Sponsor.

Objective	TASK	RATING
1	1. Prepare arc welding equipment for open-root V-groove welds.	
2	2. Make flat welds on pads and open-root V-groove joints in the 1G position.	
2	3. Make horizontal welds on pads and open-root V-groove joints in the 2G position.	
2	4. Make vertical welds on pads and open-root V-groove joints in the 3G position.	
2	5. Make overhead welds on pads and open-root V-groove joints in the 4G position.	

ANSWER KEY

<u>Answer</u>	<u>Section Reference</u>
1. b	2.3.0
2. c	2.3.0
3. b	2.4.0
4. d	3.1.0
5. c	3.1.0
6. b	3.2.0; Figure 7
7. a	3.2.0
8. c	3.3.0
9. a	4.1.0
10. c	4.1.0
11. d	4.2.0
12. b	4.2.1
13. d	4.3.1
14. b	4.3.1
15. d	4.3.2

SMAW – Open V-Groove Welds

29110-03

Performance Accreditation Tasks

OPEN-ROOT V-GROOVE WITH E6010 AND E7018 ELECTRODES IN THE FLAT POSITION

Using ⅛" and ⁵⁄₃₂" E6010 and ³⁄₃₂" E7018 electrodes, make an open-root V-groove weld on carbon steel plate in the flat position as shown.

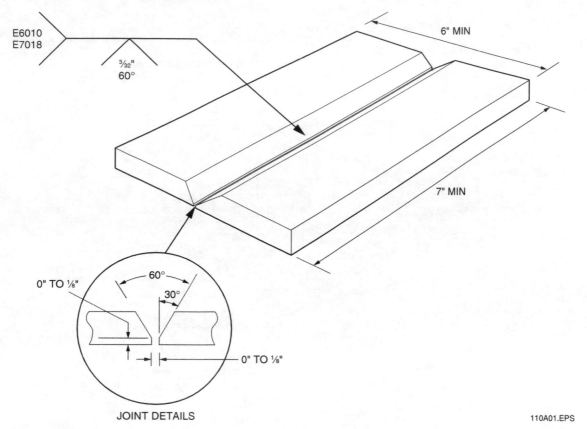

JOINT DETAILS

110A01.EPS

Criteria for Acceptance:

- Uniform rippled appearance on the bead face _____
- Craters and restarts filled to the full cross section of the weld _____
- Uniform weld size _____
- Acceptable weld profile in accordance with the *ASME Boiler and Pressure Vessel Code* _____
- Smooth transition with complete fusion at the toes of the weld _____
- Complete uniform root penetration at least flush with the base metal to a maximum buildup of ⅛" _____
- No porosity _____
- No excessive undercut _____
- No inclusions _____
- No cracks _____
- Acceptable guided bend test results _____

OPEN-ROOT V-GROOVE WITH E6010 AND E7018 ELECTRODES IN THE HORIZONTAL POSITION

Using ⅛" E6010 and ³⁄₃₂" E7018 electrodes, make an open-root V-groove weld on carbon steel plate in the horizontal position as shown.

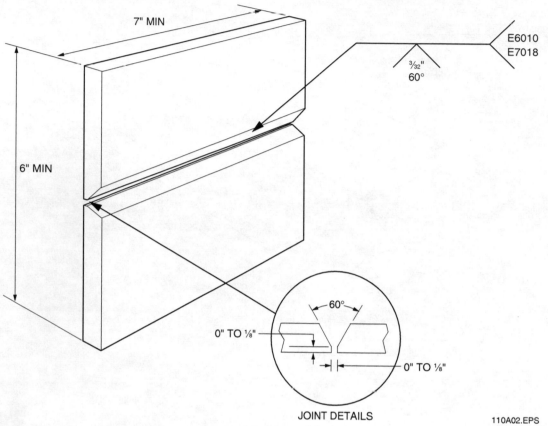

JOINT DETAILS

110A02.EPS

Criteria for Acceptance:

- Uniform rippled appearance on the bead face
- Craters and restarts filled to the full cross section of the weld
- Uniform weld size
- Acceptable weld profile in accordance with the *ASME Boiler and Pressure Vessel Code*
- Complete uniform root penetration at least flush with the base metal to a maximum buildup of ⅛"
- Smooth transition with complete fusion at the toes of the weld
- No porosity
- No excessive undercut
- No inclusions
- No cracks
- Acceptable guided bend test results

OPEN-ROOT V-GROOVE WITH E6010 AND E7018 ELECTRODES IN THE VERTICAL POSITION

Using ⅛" E6010 and ³⁄₃₂" E7018 electrodes, make an open-root V-groove weld on carbon steel plate in the vertical position as shown.

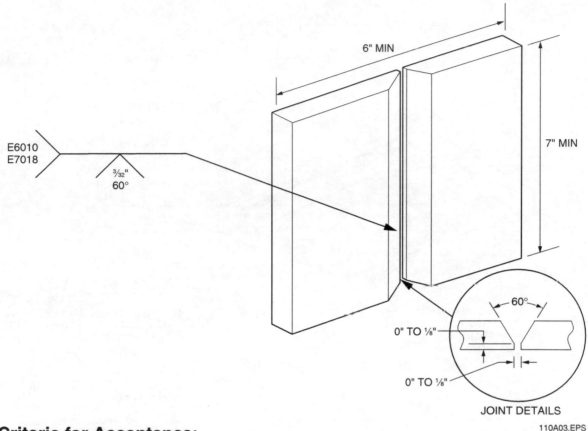

6" MIN

7" MIN

E6010
E7018

³⁄₃₂"
60°

60°

0" TO ⅛"

0" TO ⅛"

JOINT DETAILS

110A03.EPS

Criteria for Acceptance:

• Uniform rippled appearance on the bead face　　　　　　　　_____

• Craters and restarts filled to the full cross section of the weld　　_____

• Uniform weld size　　　　　　　　　　　　　　　　　　_____

• Acceptable weld profile in accordance with the
  ASME Boiler and Pressure Vessel Code　　　　　　　　_____

• Complete uniform root penetration at least flush with the
  base metal to a maximum buildup of ⅛"　　　　　　　　_____

• Smooth transition with complete fusion at the toes of the weld　_____

• No porosity　　　　　　　　　　　　　　　　　　　　_____

• No excessive undercut　　　　　　　　　　　　　　　　_____

• No inclusions　　　　　　　　　　　　　　　　　　　　_____

• No cracks　　　　　　　　　　　　　　　　　　　　　_____

• Acceptable guided bend test results　　　　　　　　　　_____

OPEN-ROOT V-GROOVE WITH E6010 AND E7018 ELECTRODES IN THE OVERHEAD POSITION

Using ⅛" E6010 and 3⁄32" E7018 electrodes, make an open-root V-groove weld on carbon steel plate in the overhead position as shown.

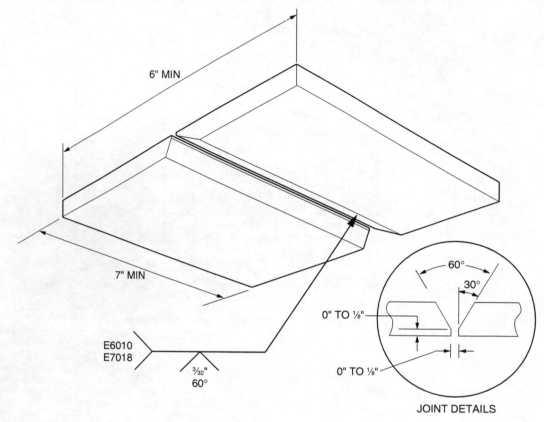

6" MIN

7" MIN

E6010
E7018

3⁄32"
60°

60°

30°

0" TO ⅛"

0" TO ⅛"

JOINT DETAILS

110A04.EPS

Criteria for Acceptance:

- Uniform rippled appearance on the bead face

- Craters and restarts filled to the full cross section of the weld

- Uniform weld size

- Acceptable weld profile in accordance with the *ASME Boiler and Pressure Vessel Code*

- Complete uniform root penetration at least flush with the base metal to a maximum buildup of ⅛"

- Smooth transition with complete fusion at the toes of the weld

- No porosity

- No excessive undercut

- No inclusions

- No cracks

- Acceptable guided bend test results

SMAW – Open-Root Pipe Welds

29111-03

Trainee Name: _____

Social Security Number: _____   **Date:** _____

_____   1.   The most common weld used
               for joining medium- and thick-
               walled pipe is the _____ weld.

               a.   overhead fillet
               b.   single-bevel groove
               c.   horizontal fillet
               d.   open-root V-groove

_____   2.   Pipe weld coupons should be
               cut from _____ diameter pipe.

               a.   ½" to 2"
               b.   1" to 2"
               c.   3" to 12"
               d.   14" to 20"

_____   3.   Pipe weld coupons should be
               cut from Schedule 40 or
               Schedule _____ pipe.

               a.   50
               b.   60
               c.   70
               d.   80

_____   4.   Each welded joint requires
               _____ coupons of the same
               size and schedule pipe.

               a.   two
               b.   three
               c.   four
               d.   five

_____   5.   When tack-welding coupons
               after the root opening is
               correct and the pipe ends are
               aligned, heavy-wall or large
               diameter pipe may require
               _____.

               a.   more and shorter tacks
               b.   longer or more tacks
               c.   a long continuous tack
               d.   double layers of tacks

_____   6.   When running E6010
               electrodes on open-root
               V-groove pipe welds, you
               should _____.

               a.   use the running keyhole
                     technique to achieve
                     proper root penetration
               b.   avoid use of a whipping
                     motion to control puddle
                     size
               c.   set the amperage in the
                     upper part of the
                     suggested range
               d.   avoid removal of slag
                     between passes

_____   7.   When the two root faces on
               the pipe edges are burned
               away by the arc, the result is
               called a _____.

               a.   bead
               b.   overlap
               c.   keyhole
               d.   root pass

8. In an open-root V-groove pipe weld, root reinforcement should be flush to within _____.

    a. ⅟₃₂"
    b. ³⁄₃₂"
    c. ⅛"
    d. ³⁄₁₆"

9. In high-pressure piping systems, root pass slag is removed by _____.

    a. scaling
    b. grinding
    c. brushing
    d. chipping

10. Open-root V-groove pipe welds can be made in the _____ positions.

    a. 1G-ROTATED, 2G, 4F, and 4G
    b. 1G-ROTATED, 2G, 5G, and 6G
    c. 1F, 2F, 4F, and 6F
    d. 1G-ROTATED, 2G, 3G, and 4G

11. When the axis of the pipe being welded is vertical, and the pipe is fixed, the weld position is called _____.

    a. 1G
    b. 2G
    c. 5G
    d. 6G

12. Pipe position 6G can vary _____ degrees from the nominal 45-degree angle.

    a. ±3
    b. ±5
    c. ±15
    d. ±20

13. Test specimens shall be cut from the four regions midway between the 12-o'clock and 3-o'clock, 3-o'clock and 6-o'clock, 6-o'clock and 9-o'clock, and 9-o'clock and 12-o'clock positions when destructive testing is required for _____ welds.

    a. 1G-ROTATED, 2G, or 6G
    b. 1G-ROTATED, 2G, or 5G
    c. 5G or 6G
    d. 1G-ROTATED or 2G

14. When running filler and cover passes on open-root V-groove pipe welds in the 2G position, use stringer beads and a slight circular or side-to-side motion to ensure tie in at the _____ of the weld bead.

    a. root
    b. face
    c. toes
    d. leg

15. When running the first filler pass on an open-root V-groove pipe weld in the 5G position, clean the face of the root pass with _____.

    a. oil
    b. water
    c. sandpaper
    d. a file or grinder

Craft: Welding

Module Number: 29111-03

Module Title: SMAW – Open-Root Pipe Welds

TRAINEE NAME: _____

TRAINEE SOCIAL SECURITY NUMBER: _____

CLASS: _____

TRAINING PROGRAM SPONSOR: _____

_____

INSTRUCTOR: _____

Rating Levels: 1. Passed: performed task.

2. Failed: did not perform task.

Recognition: When testing for the NCCER Standardized Craft Training Program, be sure to record Performance Profile testing results on Craft Training Report Form 200 and submit the results to the Training Program Sponsor.

Objective	TASK	RATING
1	1. Prepare arc welding equipment for open-root pipe welds.	
3	2. Make pipe welds in the 1G-ROTATED position.	
3	3. Make pipe welds in the 2G position.	
3	4. Make pipe welds in the 5G position.	
3	5. Make pipe welds in the 6G position.	

ANSWER KEY

<u>Answer</u>	<u>Section Reference</u>
1. d	1.0.0
2. c	2.3.0
3. d	2.3.0
4. a	2.3.0
5. b	2.3.0
6. a	2.4.0
7. c	3.1.0
8. c	3.1.0
9. b	3.1.0
10. b	3.2.0
11. b	3.2.0
12. b	3.2.0
13. c	3.2.1
14. c	4.2.0
15. d	4.3.0

SMAW – Open-Root Pipe Welds

29111-03

Performance Accreditation Tasks

OPEN-ROOT V-GROOVE PIPE WELD IN THE 1G-ROTATED POSITION

Using ⅛" E6010 electrodes for the root pass and ⅛" E7018 electrodes for the filler and cover passes, make an open-root V-groove weld on pipe in the 1G-ROTATED position as shown.

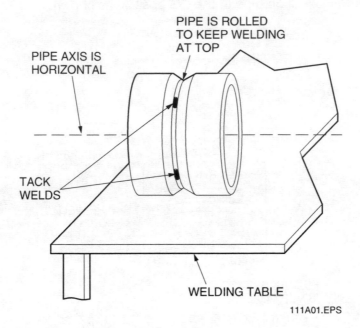

PIPE IS ROLLED TO KEEP WELDING AT TOP

PIPE AXIS IS HORIZONTAL

TACK WELDS

WELDING TABLE

111A01.EPS

Criteria for Acceptance:

- Uniform rippled appearance on the weld face   _____

- Craters and restarts filled to the full cross section of the weld   _____

- Uniform weld size   _____

- Acceptable weld profile in accordance with the
  ASME Boiler and Pressure Vessel Code, Section IX.   _____

- Smooth transition with complete fusion at the toes of the weld   _____

- Complete uniform root reinforcement at least flush with the
  inside of the pipe to a maximum buildup of ⅛"   _____

- No porosity   _____

- No excessive undercut   _____

- No inclusions   _____

- No cracks   _____

- Acceptable guided bend test results   _____

OPEN-ROOT V-GROOVE PIPE WELD
IN THE 2G POSITION (Page 1 of 2)

Using ⅛" E6010 electrodes for the root pass and ⅛" E7018 electrodes for the filler and cover passes, make an open-root V-groove weld on pipe in the 2G position as shown.

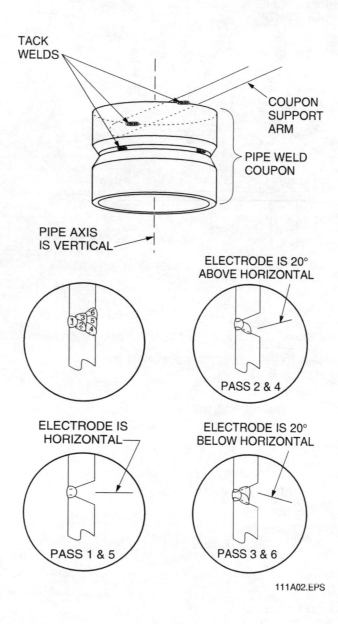

111A02.EPS

continued

OPEN-ROOT V-GROOVE PIPE WELD IN THE 2G POSITION (Page 2 of 2)

Criteria for Acceptance:

- Uniform rippled appearance on the weld face   _____

- Craters and restarts filled to the full cross section of the weld   _____

- Uniform weld size   _____

- Acceptable weld profile in accordance with the
  ASME Boiler and Pressure Vessel Code, Section IX   _____

- Smooth transition with complete fusion at the toes of the weld   _____

- Complete uniform root reinforcement at least flush with the inside
  of the pipe to a maximum buildup of ⅛"   _____

- No porosity   _____

- No excessive undercut   _____

- No inclusions   _____

- No cracks   _____

- Acceptable guided bend test results   _____

OPEN-ROOT V-GROOVE PIPE WELD IN THE 5G POSITION

Using ⅛" E6010 electrodes for the root pass and ⅛" E7018 electrodes for the filler and cover passes, make an open-root V-groove weld on pipe in the 5G position as shown.

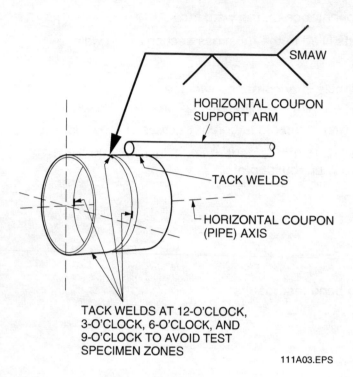

SMAW

HORIZONTAL COUPON
SUPPORT ARM

TACK WELDS

HORIZONTAL COUPON
(PIPE) AXIS

TACK WELDS AT 12-O'CLOCK,
3-O'CLOCK, 6-O'CLOCK, AND
9-O'CLOCK TO AVOID TEST
SPECIMEN ZONES

111A03.EPS

Criteria for Acceptance:

- Uniform rippled appearance on the weld face

- Craters and restarts filled to the full cross section of the weld

- Uniform weld size

- Acceptable weld profile in accordance with the
  ASME Boiler and Pressure Vessel Code, Section IX

- Smooth transition with complete fusion at the toes of the weld

- Complete uniform root reinforcement at least flush with the inside
  of the pipe to a maximum buildup of ⅛"

- No porosity

- No excessive undercut

- No inclusions

- No cracks

- Acceptable guided bend test results

OPEN-ROOT V-GROOVE PIPE WELD IN THE 6G POSITION

Using ⅛" E6010 electrodes for the root pass and ⅛" E7018 electrodes for the filler and cover passes, make an open-root V-groove weld on pipe in the 6G position as shown. If required for qualification test purposes, a restricting ring may be added to the 6G position coupon to form a 6GR position coupon.

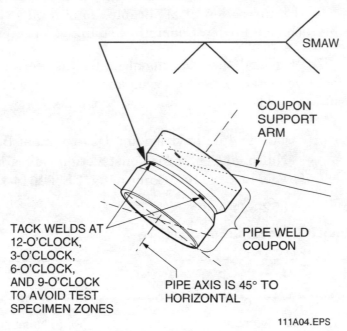

SMAW

COUPON SUPPORT ARM

PIPE WELD COUPON

TACK WELDS AT 12-O'CLOCK, 3-O'CLOCK, 6-O'CLOCK, AND 9-O'CLOCK TO AVOID TEST SPECIMEN ZONES

PIPE AXIS IS 45° TO HORIZONTAL

111A04.EPS

Criteria for Acceptance:

- Uniform rippled appearance on the weld face　　　　＿＿＿＿＿
- Craters and restarts filled to the full cross section of the weld　　　＿＿＿＿＿
- Uniform weld size　　　　＿＿＿＿＿
- Acceptable weld profile in accordance with the *ASME Boiler and Pressure Vessel Code, Section IX*　　　＿＿＿＿＿
- Smooth transition with complete fusion at the toes of the weld　　　＿＿＿＿＿
- Complete uniform root reinforcement at least flush with the inside of the pipe to a maximum buildup of ⅛"　　　＿＿＿＿＿
- No porosity　　　　＿＿＿＿＿
- No excessive undercut　　　　＿＿＿＿＿
- No inclusions　　　　＿＿＿＿＿
- No cracks　　　　＿＿＿＿＿
- Acceptable guided bend test results　　　　＿＿＿＿＿

The NCCER makes every effort to keep these textbooks up-to-date and free of technical errors. We appreciate your help in this process. If you have an idea for improving this textbook, or if you find an error, a typographical mistake, or an inaccuracy in NCCER's Contren™ textbooks, please write us, using this form or a photocopy. Be sure to include the exact module number, page number, a detailed description, and the correction, if applicable. Your input will be brought to the attention of the Technical Review Committee. Thank you for your assistance.

Instructors – If you found that additional materials were necessary in order to teach this module effectively, please let us know so that we may include them in the Equipment/Materials list in the Instructor's Guide.

Write: Curriculum Revision and Development Department
National Center for Construction Education and Research
P.O. Box 141104, Gainesville, FL 32614-1104

Fax: 352-334-0932

E-mail: curriculum@nccer.org

Craft _____ Module Name _____

Copyright Date _____ Module Number _____ Page Number(s) _____

Description _____

_____

_____

_____

_____

(Optional) Correction _____

_____

_____

(Optional) Your Name and Address _____

_____

_____